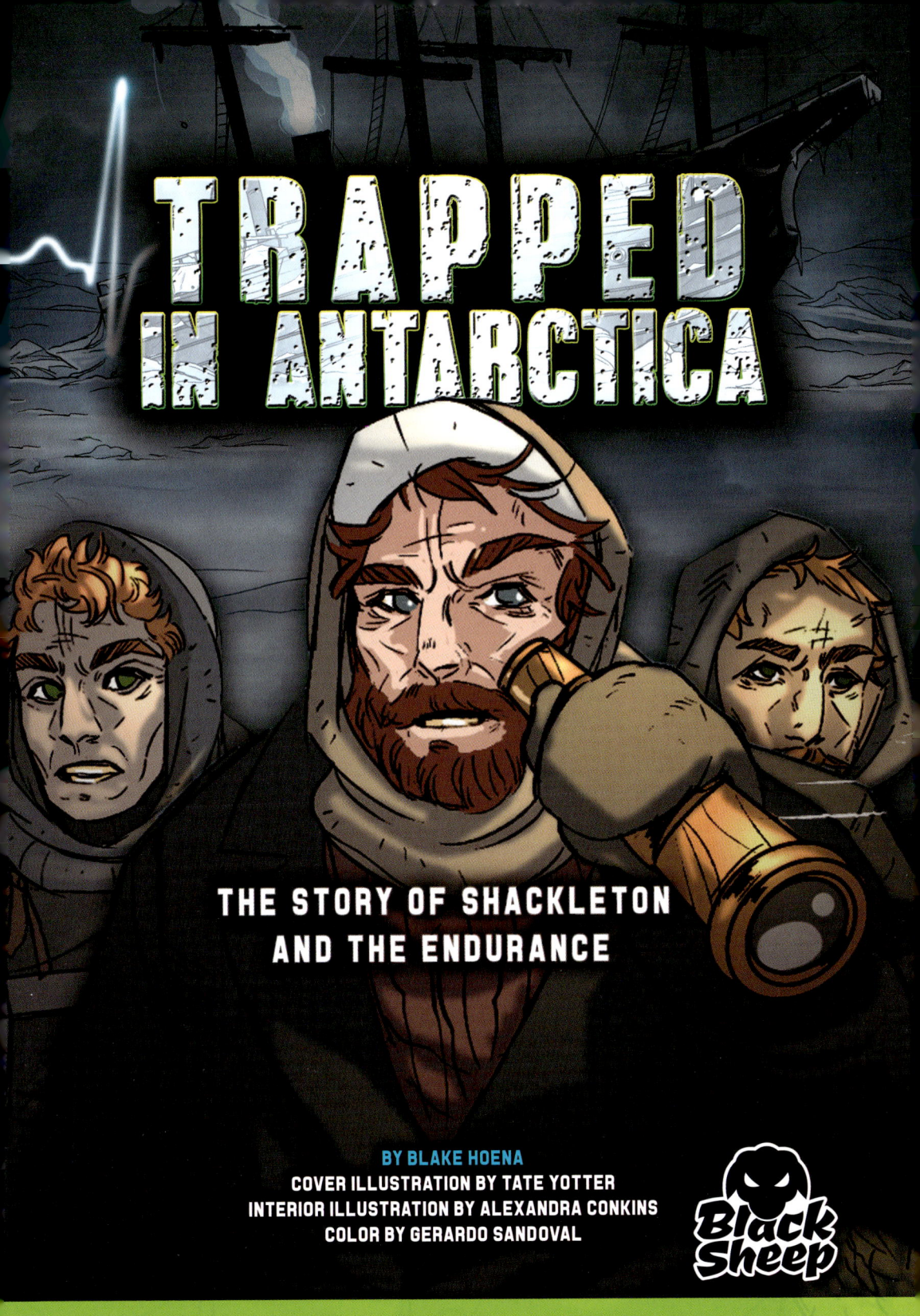

TRAPPED IN ANTARCTICA

THE STORY OF SHACKLETON AND THE ENDURANCE

BY BLAKE HOENA
COVER ILLUSTRATION BY TATE YOTTER
INTERIOR ILLUSTRATION BY ALEXANDRA CONKINS
COLOR BY GERARDO SANDOVAL

BELLWETHER MEDIA • MINNEAPOLIS, MN

STRAY FROM REGULAR READS WITH BLACK SHEEP BOOKS. FEEL A RUSH WITH EVERY READ!

This edition first published in 2022 by Bellwether Media, Inc.

No part of this publication may be reproduced in whole or in part without written permission of the publisher. For information regarding permission, write to Bellwether Media, Inc., Attention: Permissions Department, 6012 Blue Circle Drive, Minnetonka, MN 55343.

Library of Congress Cataloging-in-Publication Data

LC record for Trapped in Antarctica: The Story of Ernest Shackleton and the Endurance available at https://lccn.loc.gov/2021025032

Text copyright © 2022 by Bellwether Media, Inc. BLACK SHEEP and associated logos are trademarks and/or registered trademarks of Bellwether Media, Inc.

Editor: Betsy Rathburn   Designer: Andrea Schneider

TABLE OF CONTENTS

A NEW DREAM	4
TRAPPED IN ICE	8
RESCUE!	16
MORE ABOUT ERNEST SHACKLETON	22
GLOSSARY	23
TO LEARN MORE	24
INDEX	24

Red text identifies historical quotes.

A NEW DREAM

It is 1912. British explorer Ernest Shackleton has just learned that Roald Amundsen, a Norwegian explorer, has achieved a goal that Shackleton has had for many years.

I can't believe it. My dream... Amundsen beat me to it.

Just a few months earlier, Amundsen became the first person to reach the South Pole of Antarctica.

Shackleton is disappointed. His own two attempts to reach the South Pole ended in failure. But that does not stop him from planning an even more daring **expedition**.

I'll be the first to cross Antarctica!

The Imperial Trans-Antarctic Expedition would take Shackleton across the coldest place on Earth. In Antarctica, temperatures dip below -100 degrees Fahrenheit, and polar winds swirl up to 200 miles per hour!

To achieve his goal, Shackleton needs a hearty crew.

*Mr. Wild, you were with me on my last **venture** south. I could use you as my second-in-command.*

Sir, you know I'd trust you with my life.

4

He also needs a strong ship. He looks for one that can sail through ice-filled seas.

"What's her name?"

"I'll call it *Endurance*, Mr. Worsley, after my family motto, 'By endurance, we conquer.'"

New Zealand explorer Frank Worsley will serve as captain of the *Endurance*.

Australian photographer Frank Hurley also joins Shackleton's crew to **document** the expedition.

The *Endurance* leaves England on August 8, 1914, with Shackleton and a crew of 27 men. After sailing for several months, the ship arrives at South Georgia Island on November 5.

"The local whalers will know about ice conditions around Antarctica."

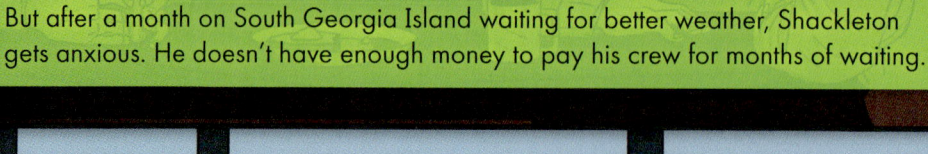

"What has the weather been like in the Weddell Sea?"

"Incredibly cold for this time of year. The ice is the worst I've ever seen."

"You'll never make it all the way to Vahsel Bay."

Vahsel Bay is where Shackleton plans to land on Antarctica. From there, he will begin his trek across the continent.

"I hope it clears soon."

But after a month on South Georgia Island waiting for better weather, Shackleton gets anxious. He doesn't have enough money to pay his crew for months of waiting.

Shackleton decides to set sail. But days later, the *Endurance* comes across an **ice floe**.

BOOM!

The ship is still 1,000 miles away from its destination.

Over the next few weeks, the *Endurance* works its way through the ice floes.
Then on the afternoon of January 18, 1915...
...the *Endurance* becomes stuck in the ice.
Shackleton has his men try to break the ship free...
It appears as though we have stuck fast for this season.
CHING! CHING! CHING!
...with no luck.

TRAPPED IN ICE

With their ship trapped in ice, the men find ways to entertain themselves. They play soccer and exercise the sled dogs.

They hunt seals and penguins to add to their food supplies.

When it gets too cold, they head below deck. They play games, read, and listen to music.

8

With the *Endurance* stuck within it, the ice floe drifts through the ocean. On February 22, 1915, it reaches the 77th **parallel**. This is the expedition's southernmost point. But then...

"The water's getting deeper, sir."

"We must be drifting north now."

"I repeat, this is the *Endurance*. We are trapped in the ice. Can anyone hear me?"

Shackleton and his crew are trapped near the bottom of the world. They try to radio for help. But they are too far from **civilization** to find any.

Shackleton admits defeat. Instead of pushing farther south, he has to make a hard choice.

"The ice floes are now carrying us away from land."

"We'll need to winter aboard the *Endurance* until the weather warms and the ice breaks apart."

For months, the expedition slowly drifts north. The men cannot control where they float.

Shackleton tries to track their movements.

Aboard the *Endurance*, the men are sheltered from the harsh weather. But the constant movements of the ice take a toll on the ship.

CRRREEEAAAKKK!

She's pretty near her end. The ship can't live in this...

What the ice gets, the ice keeps.

CRRREEEAAAKKK!

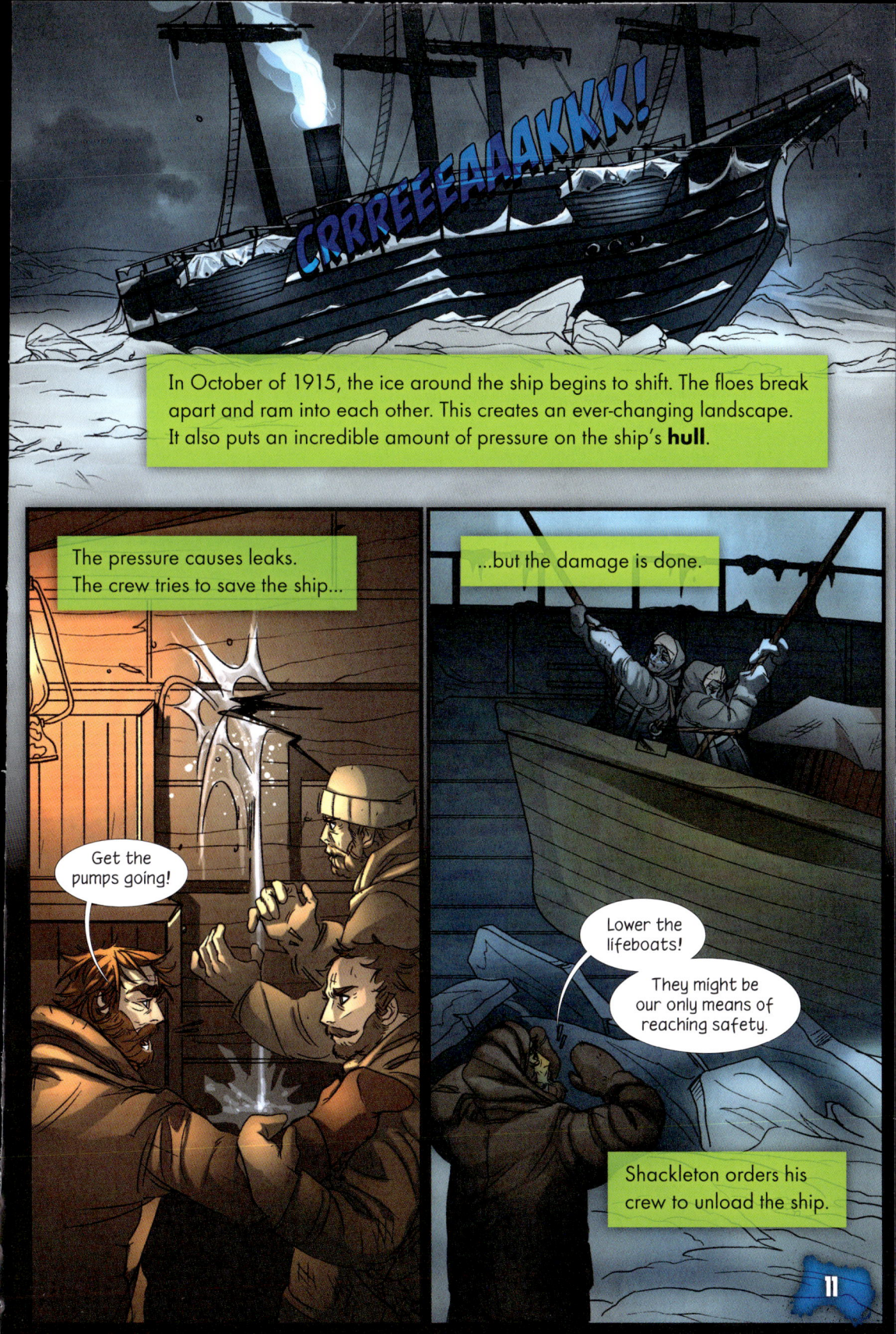

In October of 1915, the ice around the ship begins to shift. The floes break apart and ram into each other. This creates an ever-changing landscape. It also puts an incredible amount of pressure on the ship's **hull**.

The pressure causes leaks. The crew tries to save the ship...

Get the pumps going!

...but the damage is done.

Lower the lifeboats! They might be our only means of reaching safety.

Shackleton orders his crew to unload the ship.

The crew rescues most of their supplies from the *Endurance*.

Without the ship, all they have are thin tents for protection against the harsh weather.

With poor shelter and little food, Shackleton knows he must act.

"We are drifting northwest, toward the tip of the Antarctic **Peninsula**. We'll head for Paulet Island. There are food stores there."

Paulet Island is a small island about 300 miles from the *Endurance* crew's camp. They must march across the ice to get there.

They pull the lifeboats with them. The going is slow.

"We've barely covered a mile today."

"A wretched day!"

They slowly run out of supplies...

"We have nothing left to eat but seal and penguin meat."

...and give up on the march after covering little ground in three days.

The crew sets up a new camp, called Ocean Camp. There, they plan to wait until it is safe to travel again.

As they wait, the ice they are trapped on slowly drifts through the ocean.

Over the next month, the men of Ocean Camp visit the *Endurance* to recover more supplies. Each time, the ship's condition grows worse.

GRRROOOAAANNN!

"She's gone, boys."

The ship finally sinks below the ice on November 21, 1915.

The crew drifts on the ice for several months. A failed attempt to travel leads to a new camp, Patience Camp.

There, the crew settles in to wait. They endure freezing conditions, and they have little food left.

I-I-I can't feel my toes.

My sleeping bag is frozen solid.

Shackleton begins to doubt whether his crew can make it to safety. But in late March of 1916...

Land! True land and not an **iceberg**.

14

From the size of it, I believe it's Joinville Island.
That means we've drifted beyond Paulet Island and are starting to move east.
If the ice opens we could land in a day.
Best prepare the lifeboats.
The island gives the men hope that they'll soon find safety.
But as the days slip by, the island slowly disappears from sight.
If the ice doesn't open up soon, we will never reach land.

RESCUE!

On April 7, 1916, Shackleton spots a new island.

It's Elephant Island!

Elephant Island is a small, rocky piece of land about 150 miles from the northernmost tip of Antarctica. It may be the crew's last chance to reach land before they start to drift to sea.

Two days later, a crack opens in the ice.

The time for launching the boats [is] at hand!

The crew of the *Endurance* has survived on the ice for 15 months. But the most dangerous part of their journey lies ahead.

All around them, the ice floes shift and collide, threatening to destroy their fragile lifeboats.
CRUSH!
That was a close call.
They are soaked by freezing water.
Keep rowing!
I have no choice. My gloves are frozen to the oar.
And they have many miles of **treacherous** sea to go through to reach Elephant Island.
I doubt all the men will survive.

After six dangerous days at sea, the crew reaches land on April 15, 1916.

"I can't believe we made it."

"It's a good thing, too. The men wouldn't have lasted much longer."

For the first time in 497 days, the crew of the *Endurance* stands on solid ground again.

On the island, the crew hunts for seals and prepares their first hot meal in days.

Shackleton does his best to keep everyone's spirits up...

"We now have solid ground under our feet and food in our bellies."

...but he worries that they cannot survive much longer.

"Many of the men are suffering from **frostbite**."

"We'd be lucky if a handful were fit enough to continue on."

"We will starve or freeze if we stay here long."

The crew rests for about a week. Then, leaving most of the crew on Elephant Island, Shackleton sets out to sea again. He takes five crew members along with him in the strongest lifeboat, the *James Caird*.

"Hold on! It's got us!"

They sail through some of the stormiest seas on Earth.

After 17 days, the *James Caird* reaches South Georgia Island. But their journey is not over.

"The whaling station is over those mountains."

They must now hike across the island for help.

Two years after they had left the whaling station, they finally returned. They were exhausted. They had long beards, and their clothes were dirty and worn.

"Do I know you?"

"My name is Shackleton."

"Can you help us?"

Meanwhile, on Elephant Island, the men struggle to survive. They eat seal for food and upturn their lifeboats for shelter.

"They've been gone for months. Do you think they've made it?"

"If anyone can make that journey, Shackleton can."

They begin to lose hope when...

On August 30, 1916, Shackleton rescues the rest of his crew from Elephant Island.

"I have done it. Not a life lost."

Shackleton's crew had battled brutal cold and icy waters. But Shackleton's leadership helped them through all of these dangers.

In May of 1917, nearly three years after the *Endurance* departed, Shackleton returned to England. Though he never completed his goal, Shackleton's leadership made his expedition famous!

MORE ABOUT ERNEST SHACKLETON

+ Ernest Shackleton was born February 15, 1874, in Kilkea, Ireland.
+ At age 16, Shackleton joined the British Merchant Navy.
+ Shackleton served under Robert Scott on the *Discovery* Expedition to Antarctica from 1901 to 1904.
+ From 1907 to 1909, Shackleton led the *Nimrod* Expedition to Antarctica. He came within 112 miles of the South Pole.
+ Shackleton was knighted in 1909.
+ Shackleton set off on his final expedition to Antarctica in 1921. He planned to sail around the continent. But he never completed the journey. On January 5, 1922, Shackleton died while visiting South Georgia Island.

ERNEST SHACKLETON'S ROUTE

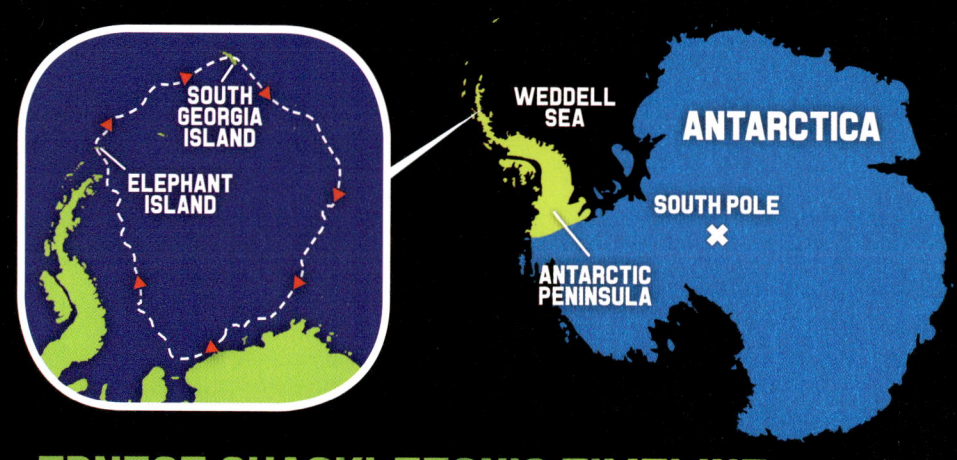

ERNEST SHACKLETON'S TIMELINE

August 8, 1914
The *Endurance* sets sail from England

January 18, 1915
The *Endurance* gets trapped in ice in the Weddell Sea

October 27, 1915
Shackleton and his crew are forced to abandon the *Endurance*

November 21, 1915
After getting crushed by the ice, the *Endurance* finally sinks

GLOSSARY

civilization—a place where people live that has stores, hospitals, and other necessities

document—to make a record of an event through such things as photographs, videos, and written reports

expedition—a trip with a purpose

frostbite—damage to the body from being frozen for a long period of time

hull—a ship's body

ice floe—a large sheet of floating ice

iceberg—a large chunk of ice floating in the water

parallel—an imaginary line that runs east and west to circle Earth; a parallel is used to measure north and south distances on the planet.

peninsula—a section of land that extends out from a larger piece of land and is almost completely surrounded by water

treacherous—very dangerous

venture—journey

April 15, 1916
Shackleton and his crew reach Elephant Island

May 10, 1916
Shackleton reaches South Georgia Island

August 30, 1916
The crew of the *Endurance* is rescued from Elephant Island

TO LEARN MORE

AT THE LIBRARY

Bell, Samantha S. *12 Extreme Survival Stories*. Mankato, Minn.: 12-Story Library, 2020.

Doeden, Matt. *Surviving Antarctica: Ernest Shackleton*. Minneapolis, Minn.: Lerner Publications, 2019.

Olson, Tod. *Lost in the Antarctic: The Doomed Voyage of the Endurance*. New York, N.Y.: Scholastic Press, 2019.

ON THE WEB

FACTSURFER

Factsurfer.com gives you a safe, fun way to find more information.

1. Go to www.factsurfer.com
2. Enter "Ernest Shackleton" into the search box and click 🔍.
3. Select your book cover to see a list of related content.

INDEX

77th parallel, 9
Amundsen, Roald, 4
Antarctica, 4, 5, 6, 12, 16
Elephant Island, 16, 17, 18, 20, 21
Endurance, 5, 6, 7, 8, 9, 10, 11, 12, 13, 16, 18, 21
England, 5, 21
historical quotes, 7, 10, 12, 13, 15, 16, 20
Hurley, Frank, 5, 7, 15
Imperial Trans-Antarctic Expedition, 4, 5, 9, 10, 21

James Caird, 20
Joinville Island, 15
Ocean Camp, 13
Patience Camp, 14
Paulet Island, 12, 15
route, 22
South Georgia Island, 5, 6, 20
timeline, 22–23
Vahsel Bay, 6
Weddell Sea, 6
Wild, Frank, 4, 15, 19
Worsley, Frank, 5, 19

24